Abdelhafid Mimouni

At the Crossroads of Metals and Life

Abdelhafid Mimouni

At the Crossroads of Metals and Life

ScienciaScripts

Imprint
Any brand names and product names mentioned in this book are subject to trademark, brand or patent protection and are trademarks or registered trademarks of their respective holders. The use of brand names, product names, common names, trade names, product descriptions etc. even without a particular marking in this work is in no way to be construed to mean that such names may be regarded as unrestricted in respect of trademark and brand protection legislation and could thus be used by anyone.

Cover image: www.ingimage.com

This book is a translation from the original published under ISBN 978-620-6-71533-7.

Publisher:
Sciencia Scripts
is a trademark of
Dodo Books Indian Ocean Ltd. and OmniScriptum S.R.L publishing group

120 High Road, East Finchley, London, N2 9ED, United Kingdom
Str. Armeneasca 28/1, office 1, Chisinau MD-2012, Republic of Moldova, Europe
Printed at: see last page
ISBN: 978-620-7-77297-1

AT THE CROSSROADS OF METALS AND LIFE

AUTHOR

Dr Abdelhafid Mimouni is an independent researcher specialising in the chemistry of bioinorganic systems, with extensive expertise in macromolecular synthesis and characterisation. He obtained his PhD in chemistry from the University of Paris XII in 1997 and a Diplôme des Etudes Approfondies des systèmes bioinorganiques from the University of Paris XI in 1993.

SUMMARY

This book provides an in-depth exploration of bioinorganic chemistry, highlighting its essential role in the understanding of biological systems and the development of medical and pharmaceutical applications. Covering the fundamental principles of inorganic chemistry relevant to biology, applications of metals in medicine, recent advances and future prospects, it highlights the continuing importance of this field for scientific and medical innovation. Each chapter offers concrete examples and relevant references, providing a comprehensive and stimulating overview of modern bioinorganic chemistry.

TABLE OF CONTENTS

INTRODUCTION 6

CHAPTER 1 8

CHAPTER 2 11

CHAPTER 3 14

CHAPTER 4 17

CHAPTER 5 20

CHAPTER 6 23

CHAPTER 7 25

GLOSSARY 29

REFERENCES 35

INTRODUCTION

Bioinorganic chemistry is an exciting interdisciplinary field that studies the interactions between inorganic compounds and biological systems. It encompasses the study of metals and their complexes, as well as organic molecules involved in essential biological processes. In this book, we explore the importance of bioinorganic chemistry in understanding fundamental biological mechanisms and in developing new medical applications.

Bioinorganic chemistry is of paramount importance in biological systems, where metals play a crucial role in a multitude of cellular processes. Transition metals, such as iron, zinc and copper, act as cofactors in enzymes and proteins, catalysing reactions essential to life. Metal ions are also involved in nutrient transport, cell signalling and metabolic regulation. Understanding the interactions between metals and biomolecules is therefore essential for deciphering complex biological mechanisms.The aim of this book is to explore in depth the fundamental principles of bioinorganic chemistry, with particular emphasis on the roles of metals in biological systems. We will also examine the medical and pharmaceutical applications of bioinorganic chemistry, particularly in the field of cancer therapy and drug design. In this context, a compound of major interest is N-acetylcysteine (NAC), a derivative of the amino acid cysteine. NAC is a powerful antioxidant and a mucolytic agent widely used in medicine. Its role in the prevention of oxidative stress and in the treatment of various ailments, notably respiratory and liver disorders, makes it an important subject of study in the field of bioinorganic chemistry. We will explore the properties of NAC and its implications for human health in detail throughout this book. In summary, this book aims to provide a comprehensive overview of bioinorganic chemistry, highlighting the importance of metals and their complexes in biological systems. We hope that this exploration will enrich readers' understanding and stimulate further research in this fascinating field.

CHAPTER 1

BASICS OF BIOINORGANIC CHEMISTRY

Bioinorganic chemistry provides the backdrop to the complex interactions between inorganic compounds and biological systems. This chapter lays the groundwork by exploring in depth the fundamental principles of inorganic chemistry relevant to this dynamic discipline.

Fundamental principles of inorganic chemistry relevant to biological systems

Understanding biological processes requires a sound knowledge of inorganic chemistry concepts. For example, studying the atomic structure of elements is essential for understanding the specificity of interactions between biomolecules and metal ions. Take the example of iron in haemoglobin The ability of iron to bind with oxygen is fundamental to the transport of this vital element in the blood. In addition, the periodic properties of elements provide a framework for understanding variations in the chemical behaviour of metals in biological environments. For example, the tendency of alkali metals to form monovalent cations is crucial to their role in regulating cell pH and maintaining ionic homeostasis. Finally, fundamental chemical reactions such as redox processes and complexation are ubiquitous in biological systems. A concrete example is photosynthesis, where the conversion of light energy into chemical energy involves processes such as redox processes involving metals such as magnesium in chlorophyll.

Chemical bonds in metal complexes

Metal complexes are key players in many biological processes. For example, the active centres of iron-containing enzymes, such as cytochrome c oxidase, use covalent and coordination bonds to catalyse redox reactions involving electron transport. In the case of zinc-based enzymes such as carboanhydrase, the zinc-ligand bond is crucial for catalysing carbon dioxide hydration reactions, an essential step in the regulation of tissue pH.

Main classes of biologically relevant inorganic compounds

Transition metals, with their flexible coordination properties, are often involved in enzyme catalysis. For example, the active site of superoxide dismutase, an enzyme essential for defence against free radicals, contains a copper and zinc active centre that catalyses the dismutation of the superoxide anion.

Alkali and alkaline earth metals are also crucial for many biological functions. For example, calcium acts as an intracellular messenger in cell signalling, regulating a multitude of cellular processes, including muscle contraction and hormone secretion. Finally, metalloids such as selenium and arsenic are integrated into certain proteins and enzymes, such as glutathione peroxidase, where they play a crucial role in protecting against oxidative damage.

CHAPTER 2

BIOCHEMISTRY OF METALS

The biochemistry of metals explores the multiple roles that these elements play in biological systems, from catalytic and structural functions to the regulatory roles vital to life.

The role of metals in biological systems

Metals perform a variety of critical functions in biological systems. For example, iron is an essential component of haemoglobin, the protein responsible for transporting oxygen in the blood. Each haemoglobin molecule contains a ferrous ion (Fe^{2+}) at the centre of its heme group, enabling oxygen to be bound to the red blood cells for transport to the tissues.Calcium plays a fundamental role in bone structure by forming complexes with phosphate to form hydroxyapatite, a substance that gives bones strength and rigidity. In addition, zinc acts as a cofactor in numerous enzymes, such as carboanhydrase, which catalyses the conversion of carbon dioxide into bicarbonate to regulate blood pH and promote the transport of carbon dioxide.Finally, magnesium acts as an essential ion in many enzymatic reactions, regulating the speed and specificity of these processes. For example, in the ATP (adenosine triphosphate) synthesis reaction, magnesium is needed to stabilise the substrate structure and facilitate energy transfer.

Metalloproteins: structure and function

Metalloproteins are proteins that contain metal ions in their structure and play vital roles in many biological processes. A notable example is catalase, an enzyme present in cells that catalyses the breakdown of hydrogen peroxide into water and oxygen. The active centre of catalase contains a heme group with a ferrous ion (Fe^{2+}), which is crucial for the catalytic reaction.Similarly, ferritin is a metalloprotein involved in the storage of iron in cells. It forms a protein-iron complex that enables iron to be stored in a soluble and safe way, ready to be released when needed for biological processes.

Metalloenzymes: catalytic mechanisms and reactions involved

Metalloenzymes are enzymes that use metal ions to catalyse specific chemical reactions. An emblematic example is photosynthesis, where metals play a central role in the conversion of light energy into chemical energy. The active centres of photosynthetic complexes contain manganese and magnesium ions that facilitate charge separation and electron transfer, enabling the production of ATP and NADPH.Another example is carbonic anhydrase, an enzyme that catalyses the reversible conversion of carbon dioxide into bicarbonate and protons. Carbonic anhydrase contains a zinc ion in its active site, which stabilises the reactants. and speeds up the reaction, facilitating the transport of carbon dioxide in the blood.

CHAPTER 3

BIOLOGICAL ANTIOXIDANTS: N-ACETYLCYSTEINE AND GLUTATHIONE

Biological antioxidants, such as N-acetylcysteine (NAC) and glutathione, play a crucial role in protecting against oxidative stress, an imbalance between the production of free radicals and the ability of biological systems to neutralise them. This chapter explores in detail the roles, mechanisms of action and medical applications of these essential antioxidants.

The role of glutathione and N-acetylcysteine in protection against oxidative stress

Glutathione (GSH) is a tripeptide found in large quantities in cells, where it acts as a powerful antioxidant and an essential cofactor for numerous enzymes. Its biosynthesis begins with the conjugation of glutamate, cysteine and glycine by γ-glutamylcysteine synthetase (γ- GCS) to form γ-glutamylcysteine, which is then converted to GSH by glutathione synthetase.N-acetylcysteine (NAC) is a precursor of cysteine, a key amino acid in the synthesis of glutathione. NAC is rapidly metabolised in cells, where it is hydrolysed to release cysteine. This cysteine is then used in glutathione biosynthesis, increasing intracellular GSH levels. In addition, NAC acts as a thiol group (-SH) donor, offering direct protection against free radicals and oxidative damage.

Mechanisms of action of antioxidants in cells and tissues

Glutathione and N-acetylcysteine protect cells and tissues against oxidative damage by neutralising free radicals and recycling oxidised antioxidants. Glutathione neutralises hydroxyl radicals and lipid peroxides by forming disulphide bonds (-S-S-) with them, while NAC provides thiol groups (-SH) that reduce reactive species. In addition to their direct antioxidant activity, glutathione and NAC regulate the expression of genes involved in the response to oxidative stress. For example, glutathione activates the transcription factor Nrf2 (factor 2 linked to nuclear factor erythroid 2), which induces the expression of genes encoding antioxidant enzymes and detoxification proteins.

Medical and therapeutic applications of N-acetylcysteine and glutathione

N-acetylcysteine and glutathione have a variety of medical applications due to their antioxidant, anti-inflammatory and mucolytic properties. NAC is widely used in the treatment of paracetamol poisoning by increasing hepatic glutathione production. NAC is also effective in the treatment of respiratory disorders, such as chronic obstructive pulmonary disease (COPD), due to its mucolytic properties, which encourage expectoration of mucus. Glutathione also has therapeutic applications in the treatment of neurodegenerative diseases, cardiovascular disease and cancer. From Studies have shown that reduced glutathione levels are associated with many chronic diseases, underlining the importance of maintaining an optimal glutathione balance for human health.

CHAPTER 4

TRANSPORT AND STORAGE OF METALS

Metal transport and storage in organisms are complex and highly regulated processes, essential for maintaining metal homeostasis and supporting many biological functions. This chapter explores in depth the mechanisms of metal transport, storage and regulation in cells and organisms.

Transport of metals in organisms

Metals are transported across cell membranes by a variety of specialised proteins, including transporters and chelators. A recent example is the recently discovered ZIP8 transporter, which is involved in the transport of zinc across the plasma membrane and in the regulation of intracellular zinc concentration. ZIP8 plays a crucial role in numerous biological processes, such as cell growth, signalling and zinc homeostasis. In addition, chelators, such as metallothionein, are cysteine-rich proteins that bind to metals and transport them across cell membranes. Metallothionein protects cells against the toxic effects of heavy metals and regulates intracellular concentrations of zinc, copper and other essential metals.

Mechanisms for storing metals in cells

Metals are stored in cells in the form of complexes with specific storage proteins or in the form of intracellular granules. An innovative example is ferritin, an iron storage protein that form complexes with iron and store it as ferricity (Fe^{3+}) in the cells. Ferritins play a crucial role in regulating intracellular iron concentration and protecting against oxidative stress.

In addition, cells can store metals in the form of intracellular granules, as is the case for zinc, which is stored in the form of zinc granules in lysosomal vesicles. These zinc granules act as zinc stores, providing a constant supply of zinc for biological processes when required.

Metal homeostasis: regulation of metal concentrations in cells and organisms

Metal homeostasis is maintained by a delicate balance between the processes of transport, storage and elimination of metals. Recent examples show that dysfunctions in these processes can lead to serious diseases, such as Wilson's disease, characterised by an excessive accumulation of copper in tissues. Organisms have developed sophisticated mechanisms for regulating metal concentrations in cells and tissues. One example is metallothionein proteins, whose expression is regulated by intracellular metal levels and environmental stressors. These proteins act as metal sensors, adjusting intracellular metal concentration to maintain homeostasis.

CHAPTER 5

TOXICOLOGY OF METALS

Metal toxicology studies the adverse effects of metals on human health and the environment, as well as the mechanisms underlying their toxicity. This chapter explores the sources of exposure to toxic metals, the associated mechanisms of toxicity and the consequences for human health and the environment.

Sources of exposure to toxic metals

Sources of exposure to toxic metals are varied and ubiquitous in our environment. The main sources include industrial emissions, toxic waste, pesticides, consumer products (such as cosmetics and toys), and contaminated water and food. For example, lead is often present in old paint, lead pipes and soil contaminated by industrial activities.

Mechanisms of metal toxicity

Toxic metals exert their harmful effect by disrupting vital cellular processes through a number of mechanisms, including:

• Interaction with proteins: Toxic metals can bind to cellular proteins, disrupting their structure and function. For example, mercury can bind to sulphhydrylated proteins, inhibiting essential enzymes.

• Generation of free radicals: Certain toxic metals, such as lead and cadmium, can induce the formation of free radicals, leading to oxidative damage to cell membranes, DNA and proteins.

• Disruption of cellular processes: Toxic metals can disrupt cellular processes by interfering with nutrient transport, altering cell metabolism and disrupting cell signalling pathways.

Consequences for human and environmental health

The consequences of exposure to toxic metals for human health are serious and varied. Short-term effects can include gastrointestinal symptoms, neurological disorders, kidney and liver damage, while long-term effects can include cancer, cardiovascular disease, chronic neurological disorders and developmental alterations in children.

Environmentally, exposure to toxic metals can have devastating effects on ecosystems, leading to contamination of soil, groundwater and watercourses. This can have an impact on biodiversity, the quality of drinking water and agricultural activities.

CHAPTER 6

MEDICAL AND PHARMACEUTICAL APPLICATIONS OF METALS

Metals play a crucial role in the medical and pharmaceutical fields, where their diversity and versatility are used to develop new therapies and innovative medicines. This chapter explores in detail the applications of metals in these fields, highlighting recent advances and future prospects.

Use of metal complexes in chemotherapy

Metal complexes, such as cisplatin, have revolutionised cancer treatment by acting directly on the DNA of cancer cells. Cisplatin, for example, forms covalent bonds with DNA, disrupting its replication and causing cancer cells to die by apoptosis. More recently, ruthenium and iridium complexes have emerged as promising alternatives to cisplatin, offering increased efficacy and reducing toxic side effects.

Applications of metal catalysts in drug synthesis

Metal catalysts are essential in drug synthesis, enabling selective and efficient chemical reactions. For example, palladium complexes are widely used in cross-coupling reactions to form carbon-carbon and carbon-heteroatom bonds, a key step in the manufacture of many drugs, including anti-cancer kinase inhibitors and protease inhibitors for HIV.

Design of bioinorganic drugs targeting specific biological processes

Bioinorganic drugs exploit the properties of metal complexes to specifically target relevant biological processes. For example, platinum complexes functionalised with ligands targeting cancer cells increase the selectivity of anti-cancer treatments, while ruthenium complexes targeting the mitochondria of cancer cells induce selective apoptosis. These approaches make it possible to improve the efficacy of treatments while minimising undesirable side effects.

CHAPTER 7

RECENT ADVANCES AND FUTURE PROSPECTS

This chapter explores recent advances in the field of bioinorganic chemistry, highlighting new experimental approaches, developments in the design of bioinorganic catalysts and drugs, and prospects for future research.

New experimental approaches in bioinorganic chemistry

Recent advances in spectroscopy, crystallography and computational modelling have revolutionised bioinorganic chemistry.

For example, high-field nuclear magnetic resonance (NMR) spectroscopy now enables detailed analysis of complex molecular structures and dynamics, opening up new perspectives for the study of metalloprotein-ligand interactions.

Recent developments in the design of catalysts and bioinorganic drugs

The design of catalysts and bioinorganic drugs has been greatly influenced by advances in coordination chemistry and structural biology. For example, the development of transition metal catalysts for selective and durable organic synthesis reactions has been intensified, enabling the efficient manufacture of complex pharmaceutical compounds.

In the field of bioinorganic drugs, significant progress has been made in the design of metal complexes for the treatment of cancer and other diseases. Examples include the use of nanoparticles functionalized gold to selectively target cancer cells and theranostic agents that combine diagnostic and therapeutic properties.

Prospects for future research in bioinorganic chemistry

The future of bioinorganic chemistry is bright, with exciting prospects for new discoveries and applications. Emerging areas of research include molecular bioimaging, bioinspired catalysis and personalised medicine. Advances in these areas will require interdisciplinary collaboration between chemists, biologists, physicists and engineers to meet the complex challenges of science and medicine.

Bioinorganic chemistry is a fascinating field that continues to inform our understanding of biological systems and drive new innovative applications. In this book, we have explored in depth the fundamental principles of inorganic chemistry relevant to biological systems, the essential roles of metals in biological processes, the medical and pharmaceutical applications of metals, and recent advances and future prospects in this exciting field.

We have seen how metals play a crucial role in many biological functions, acting as catalysts, regulators and structures in a wide range of cellular processes. Their use in the design of drugs, catalysts and functional materials offers innovative solutions to challenges in medicine, chemistry and biology.

In summary, bioinorganic chemistry remains a dynamic and constantly evolving field, offering immense potential for future discoveries and applications. Its importance continues to grow in our quest to understand life at the molecular level and to develop advanced technologies that benefit society. We hope that this book has helped to broaden your appreciation of this fascinating field and to stimulate your interest in its many possibilities.

GLOSSARY

- Inorganic chemistry: Branch of chemistry that studies the compounds and reactions of elements other than carbon.
- Biomolecules: Molecules present in living organisms, such as proteins, nucleic acids, lipids and carbohydrates.
- Metal complexes: Molecules containing one or more metal ions bound to ligands, forming stable structures.
- Ionic bond: A type of chemical bond in which atoms transfer electrons to form positively and negatively charged ions, which then attract each other.
- Covalent bond: Type of chemical bond in which two atoms share a pair of electrons.
- Coordination bond: Type of bond in metal complexes where a metal ion is linked to ligands by electron pairs.
- Oxidation-reduction reaction: Chemical reaction in which electrons are transferred between the reactants.
- Transition metals: Chemical elements in groups 3 to 12 of the

periodic table, characterised by their ability to form stable complexes with ligands.

• Alkali metals and alkaline earth metals: Groups of alkali metals (group 1) and alkaline earth metals (group 2) in the periodic table, respectively.

• Metalloids: Chemical elements located on the border between metals and non-metals in the periodic table, sharing certain properties of both.

• Metalloproteins: Proteins that contain metal ions in their structure and play a crucial role in many biological processes.

• Metalloenzymes: Enzymes which contain metal ions in their active site and which catalyse specific chemical reactions.

• Cofactor: Non-protein molecule required for the catalytic activity of an enzyme.

• Active centre: The part of an enzyme where chemical reactions take place, generally consisting of a substrate-binding site and residues essential for catalysis.

• Metal ion: A metal atom that has lost or gained electrons, thereby acquiring an electric charge.

• Catalysis: Acceleration of a chemical reaction by the presence of a substance, called a catalyst, which is not consumed during the reaction.

• Haemoglobin: Protein present in the red blood cells, responsible for transporting oxygen from the lungs to the body's tissues.

• Ferritin: An iron storage protein present in cells, which plays a crucial role in regulating intracellular iron concentration.

• Carbonic anhydrase: Enzyme which catalyses the reversible conversion of carbon dioxide into bicarbonate and protons, essential for the transport of carbon dioxide in the blood.

• γ-glutamylcysteine synthetase (γ-GCS): Enzyme which catalyses the first stage of glutathione biosynthesis by conjugating glutamate and cysteine to form γ-glutamylcysteine.

• Antioxidant recycling: Process by which antioxidants regenerate other antioxidants by neutralising the oxidised forms of these molecules.

• Nrf2 (nuclear factor erythroid 2-related factor 2): Glutathione-activated transcription factor that induces the expression of genes involved in the response to oxidative stress.

• ZIP8: Zinc transport protein 8, involved in the transport of zinc across the plasma membrane.

• Metallothionein: Cysteine-rich protein which binds metals and transports them across cell membranes.

• Ferritin: Iron storage protein which forms complexes with iron and stores it in the form of ferricity (Fe^3+) in the cells.

• Free radicals: Highly reactive molecules containing one or more unpaired electrons, which can cause cell damage.

• Acute toxicity: Adverse health effects occurring shortly after a single or short-term exposure to a toxic substance.

• Chronic toxicity: adverse health effects resulting from prolonged or repeated exposure to a toxic substance.

• Apoptosis: The programmed process of cell death.

• Cross-coupling: Chemical reaction in which two organic fragments are combined via a covalent bond formed by a metal catalyst.

• Protease inhibitor: Drug which blocks the action of a protease, an enzyme necessary for viral replication.

• Nanoparticle: Particle of nanometric size, often used as a vector for targeted drug delivery.

• Prodrug: An inactive form of a drug that is converted into an active form in the body.

• Antibiotic resistance: The ability of bacteria to resist the effects of antibiotics, making infections more difficult to treat.

• Antivirals: Drugs used to treat viral infections by inhibiting virus replication or stimulating the host's immune response.

• Molecular bioimaging: Technique for visualising and studying biological processes at the molecular level using specific molecular probes.

• Bioinspired catalysis: Using biological principles to design artificial catalysts capable of carrying out chemical reactions with an efficiency and selectivity similar to those of biological enzymes.

• Personalised medicine: A medical approach that uses information about an individual's genetic, molecular and environmental profile to personalise medical and preventive treatments.

REFERENCES

Berg, J. M., Tymoczko, J. L., & Stryer, L. (2002). Biochemistry (5th ed.). W H Freeman.

Lippard, S. J., & Berg, J. M. (1994). Principles of Bioinorganic Chemistry. University Science Books.

Cotton, F. A., & Wilkinson, G. (1988). Advanced Inorganic Chemistry (5th ed.). John Wiley & Sons.

Que, L. Jr. and Tolman, W. B. (Eds.). (2013). Bioinspired Catalysis: Metal-Sulfur Complexes. Springer.

Bertini, I., Gray, H. B., Stiefel, E. I., & Valentine, J. S. (2007). Biological Inorganic Chemistry: Structure and Reactivity. University Science Books.

Kambe, T., & Andrews, G. K. (2009). Novel proteolytic processing of the ectodomain of the zinc transporter ZIP8 (SLC39A8) during zinc deficiency is inhibited by acrodermatitis enteropathica mutations. Molecular and Cellular Biology, 29(1), 129-139.

Robinson, N. J., & Winge, D. R. (2010). Copper metallochaperones. Annual Review of Biochemistry, 79(1), 537-562.

Liuzzi, J. P., & Cousins, R. J. (2004). Mammalian zinc transporters. Annual Review of Nutrition, 24(1), 151-172.

Krezel, A., & Maret, W. (2007). Thionein/metallothionein control Zn(II) availability and the activity of enzymes. Journal of Biological Inorganic Chemistry, 12(8), 1011-1018.

Nordberg, G. F., Fowler, B. A., & Nordberg, M. (2014). Handbook on the Toxicology of Metals. Academic Press.

Clarkson, T. W., Magos, L., & Myers, G. J. (2003). The Toxicology of Mercury

-Current Exposures and Clinical Manifestations. The New England Journal of Medicine, 349(18), 1731-1737.

Agency for Toxic Substances and Disease Registry (2020). Toxicological Profile for Lead. U.S. Department of Health and Human Services.

Johnstone, T. C., Suntharalingam, K., & Lippard, S. J. (2016).

The Next Generation of Platinum Drugs: Targeted Pt(II) Agents, Nanoparticle Delivery, and Pt(IV) Prodrugs. Chemical Reviews, 116(5), 3436-3486.

Hartwig, J. F. (2012). Evolution of C-C and C-X Bond-Forming Reactions Using Transition Metal Complexes. Science, 337(6095), 1625-1630.

Meggers, E. (2015). Exploring Biologically Relevant Chemical Space with Metal Complexes. Accounts of Chemical Research, 48(3), 756-768.

Johnstone, T. C., Suntharalingam, K., & Lippard, S. J. (2016). The Next Generation of Platinum Drugs: Targeted Pt(II) Agents, Nanoparticle Delivery, and Pt(IV) Prodrugs. Chemical Reviews, 116(5), 3436-3486.

Hartwig, J. F. (2012). Evolution of C-C and C-X Bond-Forming Reactions Using Transition Metal Complexes. Science, 337(6095), 1625-1630.

Meggers, E. (2015). Exploring Biologically Relevant Chemical Space with Metal Complexes. Accounts of Chemical Research, 48(3), 756-768.

Que, L., & Tolman, W. B. (Eds.). (2014). Comprehensive Coordination Chemistry II: From Biology to Nanotechnology. Elsevier.

Bochmann, M. (Ed.). (2018). Catalysis: Concepts and Green Applications. Wiley.

Rosenberg, B., & Lippert, B. (Eds.). (2018). Metallo-Drugs: Development and Action of Anticancer Agents. De Gruyter.

R. K. Poole, "The emergence of transition metal biology and the rise of inorganic biochemistry," Nat. Chem. Biol. vol. 4, no 5, p. 263-267, May 2008.

E. R. Stadtman, "Metal ion-catalyzed oxidation of proteins: Biochemical mechanism and biological consequences," Free Radic. Biol. Med, vol. 9, no 4, p. 315-325, 1990.

A. A. Qamar et al, "Therapeutic uses of N-acetylcysteine (NAC) in medicine," Open Respir. Med. J., vol. 6, pp. 124-132, 2012.

Valentine, J. S., Bertini, I., Gray, H. B., & Stiefel, E. I. (Eds.). (2006). Biological Inorganic Chemistry: Structure and

Reactivity. University Science Books.

Solomon, E. I., Baldwin, M. J., & Lowery, M. D. (1992). Electronic Structures of Active Sites in Copper Proteins: Contributions to Reactivity. Chemical Reviews, 92(3), 521-542.

Kretsinger, R. H., & Uversky, V. N. (Eds.). (2012). The Structural Basis of Biological Function. Springer Science & Business Media.

Holm, R. H., Kennepohl, P., & Solomon, E. I. (1996). Structural and Functional Aspects of Metal Sites in Biology. Chemical Reviews, 96(7), 2239-2314.

Sigel, A., & Sigel, H. (Eds.). (2002). Metal Ions in Biological Systems. Marcel Dekker.

Meister, A., & Anderson, M. E. (1983). Glutathione. Annual Review of Biochemistry, 52(1), 711-760.

De Flora, S., & Grassi, C. (1996). Glutathione in pharmacology and clinical medicine. Pharmacology & Therapeutics, 67(1), 146-165.

Sadowska, A. M., & Manuel-Y-Keenoy, B. (2003). Antioxidant and anti inflammatory efficacy of NAC in the treatment of COPD: Discordant in vitro and in vivo dose-effects: A review. Pulmonary Pharmacology & Therapeutics, 16(3), 215-224.

Rushworth, G. F., Megson, I. L., & Llewellyn, D. H. (2014). Glutathione: biosynthesis, regeneration and role in redox signaling and oxidative stress. Journal of Biological Chemistry, 289(13), 7114-7121.

MIX
Papier aus verantwortungsvollen Quellen
Paper from responsible sources
FSC® C105338

Printed by Books on Demand GmbH, Norderstedt / Germany